SCIENTIFIC AMERICAN EDUCATIONAL PUBLISHING

WHAT A ZOO!

10 FUN EXPERIMENTS ABOUT ANIMALS

BRING SCIENCE HOME

Published in 2024 by The Rosen Publishing Group, Inc.
2544 Clinton Street, Buffalo, NY 14224

Contains material from Scientific American , a division of Springer Nature America, Inc., reprinted by permission as well as original material from The Rosen Publishing Group.

First Edition

Editor: Kristen Rajczak Nelson
Designer: Rachel Rising

All illustrations by Continuum Content Solutions

Activity on p. 7 by Science Buddies (October 2, 2014); p. 11 by Science Buddies (March 20, 2014); p. 17 by Science Buddies (January 16, 2014); p. 23 by Science Buddies (October 3, 2013); p. 29 by Science Buddies (August 1, 2013); p. 35 by Daisy Yuhas (October 11, 2012); p. 41 by Science Buddies (June 28, 2012); p. 47 by Science Buddies (May 10, 2012); p. 52 by Science Buddies (May 3, 2012); p. 57 by Science Buddies (December 9, 2011).

Photo Credits:pp. 3, 4, 11, 14, 29, 33, 41, 44, 47, 50 cve iv/Shutterstock.com; pp. 4, 7, 11, 17, 23, 29, 35, 41, 47, 52, 57 Anna Frajtova/Shutterstock.com.

Cataloging-in-Publication Data
Names: Scientific American, Inc.
Title: What a zoo!: 10 fun experiments about animals / edited by the Scientific American Editors.
Description: New York : Scientific American Educational Publishing, 2024. | Series: Bring science home | Includes glossary and index.
Identifiers: ISBN 9781725349308 (pbk.) | ISBN 9781725349315 (library bound) | ISBN 9781725349322 (ebook)
Subjects: LCSH: Zoology--Experiments--Juvenile literature. | Animals--Experiments--Juvenile literature. | Zoology projects--Juvenile literature.
Classification: LCC QL52.6 W43 2024 | DDC 590.78--dc23

CPSIA Compliance Information: Batch #CSSA24. For further information contact Rosen Publishing, New York, New York at 1-800-237-9932.

CONTENTS

THESE ACTIVITIES INCLUDE SCIENCE FAIR PROJECT IDEAS.

INTRODUCTION

From biodiversity to blubber, animal science is fascinating to study! Whether you love bugs or puppies, the activities and experiments in this book will engage you with the many facets of the animal kingdom. You'll investigate how to best repel ants, go bird-watching, practice "hunting," and much more!

Projects marked with include a section called Science Fair Project Ideas. These ideas can help you develop your own original science fair project. Science fair judges tend to reward creative thought and imagination, and it helps if you are really interested in your project. You will also need to follow the scientific method. See page 61 for more information about that.

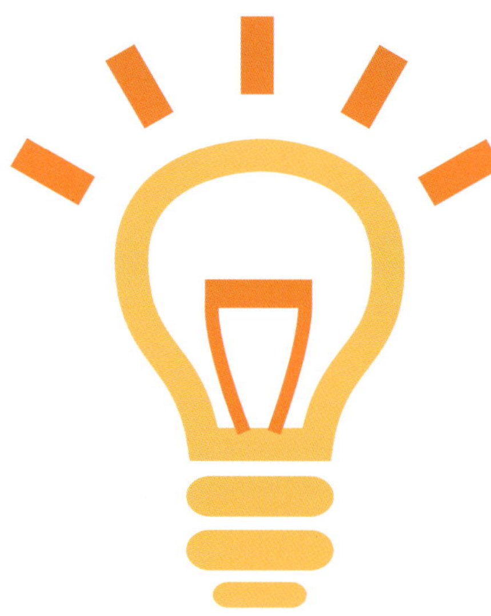

Systems Science

How Biodiverse Is Your Backyard?

EXPLORE YOUR BACKYARD OR A LOCAL PARK TO DISCOVER THE MANY KINDS OF ANIMALS THAT MAY LIVE THERE!

PROJECT TIME

45 minutes

KEY CONCEPTS

Biology
Ecology
Taxonomy
Animals
Biodiversity

Have you ever wondered how many different types of animals live near you—in your backyard or a local park? And we're not just talking about deer or even birds. Animals come in all shapes and sizes—from big bears down to the tiniest bugs—each a unique part of the amazing diversity of life. The differences among animals can help us classify animals into different groups. One way people classify animals is by their phyla. Do you know which phylum you belong to? What about a garden spider? In this science activity, you'll investigate the biodiversity of the animal life around your home and try to figure out to which phylum most of the animals belong.

BACKGROUND

From the largest elephant to the tiniest water flea, all animals have unique characteristics. Based on their shared characteristics, animals are put into categories called "phyla." (A phylum is specifically a taxonomic ranking that appears between kingdom and class in the traditional hierarchy of ranks.) There are perhaps as many as 35 different phyla within the animal kingdom alone, but most are very uncommon. Scientists recognize eight major phyla to describe most common animals: Porifera, Cnidaria, Platyhelminthes, Annelida, Mollusca, Arthropoda, Echinodermata, and Chordata. These names might sound odd at first, but once you find real-world examples close to home, some should start to become much more familiar!

Depending on where you live, the phyla you encounter most frequently may be Chordata, Arthropoda, Mollusca, and Annelida. Chordata is the most well-known group, even though the number of individuals it contains is quite small. Have you guessed which one it is? It is the phylum that includes *all* of the vertebrates, and thus is the phylum we belong to—along with most of our pets (dogs, cats, rabbits, rats, fish, frogs, salamanders), farm animals (cows, pigs, sheep, chickens), and wild animals (zebras, lions, tigers, pandas, giraffes, polar bears, etcetera). Arthropods are invertebrates and include insects (such as beetles, butterflies, flies, bees, wasps, grasshoppers, and ants) as well as centipedes, millipedes, spiders, shrimp, crayfish, pill bugs, and many others. Mollusca includes snails, slugs, bivalves (such as clams and oysters), squids, and octopuses. Annelids are segmented worms and include earthworms (which are used for compost and gardening) and aquatic worms. What phylum do you think is most common around you?

MATERIALS

- Sheet of paper
- Pencil or pen
- Digital camera (optional)
- Magnifying glass (optional)
- An observation location

PREPARATION

- Choose your observation location. It should be a place you like to explore and where you think you might find a diversity of organisms. It could be a backyard, park, community garden, open field, lake, stream, etcetera, where you are allowed to poke around.

- Pick a good time to do your investigation and bring along an adult supervisor if need be.

PROCEDURE

- Bring your sheet of paper and pen or pencil (and digital camera and magnifying glass, if desired, along with an adult, if necessary) to the observation location you decided on.

- Pick a small part of the location to investigate first. For example, it could be a pile of dead leaves, some shady rocks, a patch of lawn, old logs, dry weeds, a tree, etcetera. (These are all different microhabitats.) You will be able to choose additional microhabitats to investigate later.

- Carefully look for animals in the microhabitat you picked. You may want to turn over rocks or logs to look for animals underneath them. On your sheet of paper, write down what type of animals you see. (For example, maybe you see a bird, earwigs, or spiders.) Do not worry if you cannot identify something right away—you can write down a description, make a drawing, or take a picture and try to identify it later. *What animals do you see? What kind of microhabitat are they living in? (Is it dry or damp? Does it have plants, rocks, etcetera?)*

- Move on to another microhabitat and similarly investigate it. *What animals do you see in this area? Do any of the animals look like different species of types you saw already? What is this microhabitat like and how is it similar or different from the previous one?*

- Repeat your observations in other microhabitats—trying to find ones that are very different (sunny versus shady or dry versus damp, for example) and some similar (both shady or both dry, for example). Try to explore at least five different microhabitats at your observation location, if possible. *Do some animals seem to prefer certain types of microhabitats (such as damp ones or dry ones) more than other animals?*

- Once you're done investigating the observation location, try to categorize the animals you found by their phyla. Depending on what your observation site is, the phyla you are most likely to encounter are Chordata, Arthropoda, Mollusca, and Annelida. Remember that chordates include all vertebrates (including humans, dogs, cats, frogs, fish, cows, birds, etcetera); arthropods include insects, centipedes, millipedes, shrimp, pill bugs, etcetera; mollusks include snails and slugs; and annelids include earthworms.

- *For which phylum did you find the most kinds of animals (that is, greatest number of different species)? Which phylum was the most difficult to find? Were there some phyla that you couldn't find at all? How diverse do you think animals are in your region? Did you consistently find certain animals in a specific type of microhabitat?*

ExtRA

In this activity, you investigated a single observation location with different microhabitats, but you could also try surveying two or more separate locations and comparing the microhabitats within them. *How do your local parks compare with one another? If you live near water, how does a shoreline compare with a park's grass and trees? How does the species' diversity change?* (**Tip**: If you investigate a marine location, you'll want to do some background research on other common phyla that might live there.)

ExtRA

Make a species map to figure out how different animals are distributed in your observation location. To do this, first make a map of your observation location, marking locations of structures such as your house, a patio, shed, trees, etcetera, and make several copies of the map. Then repeat this activity but this time investigate the entire location and use a different copy of the map for each different type of organism you find, coloring in the map where you find that type of organism. *What types of areas do certain species seem to prefer? Which areas have the most different kinds of species (or most biodiversity) and which have the fewest (or least biodiversity)?*

EXTRA

What invasive organisms are in your area? What are your local populations of ladybugs, ants, bees, squirrels, birds, and other wildlife animals like? Your wildlife observations can be helpful for several different citizen science projects. See if you can find any that you can contribute your investigations to. Check out: the Citizen Science Association (citizenscience.org), Project Squirrel (www.projectsquirrel.org), or Wildlife Sightings (www.wildlifesightings.net).

OBSERVATIONS AND RESULTS

Did you mostly find arthropods at your observation location?

The types of animals you saw depends on your observation location. But for most backyards, parks, gardens, fields, etcetera, the phylum that the majority of the different animals belongs to is Arthropoda. Some of the most common arthropods you would see in North America include various flies, beetles (including ladybugs), spiders, roly-polies, grasshoppers, centipedes, millipedes, earwigs, ants, butterflies, wasps, crickets, and many others. You might also have seen snails and slugs (which belong to the Mollusca phylum). If you dug around in the soil, you may have uncovered earthworms (part of the Annelida phylum). And if you counted yourself in your survey, then you also found at least one member of the phylum Chordata! Other common chordates you may have seen include squirrels, birds, dogs, and cats.

CLEANUP

Wash your hands with soap and water. Return all your materials to where you found them.

Springtime Science

What's Home Sweet Home to a Bug?

DO PILL BUGS OR SOW BUGS PREFER A WET OR DRY HABITAT? TRY THIS ACTIVITY TO FIND OUT!

Have you ever wondered how an animal chooses where to live? With the arrival of spring, animals in many areas emerge from their winter hiding places to search for a place to live until next winter. One group of critters that can usually be found just about anywhere on land includes the familiar sow bugs and pill bugs. As they come out to enjoy the warmer weather, what type of environment do you think they'll seek? In this science activity, you'll answer part of this question using your own local pill bugs or sow bugs, so get ready to build a habitat and do some bug hunting!

PROJECT TIME

1 hour

KEY CONCEPTS

Animals
Environments
Crustaceans
Adaptations

BACKGROUND

You can often find sow bugs and pill bugs in damp, dark places, like the soil under rocks or decaying wood. These insects are both crustaceans (specifically, a type called isopods). Other crustaceans include shrimp, lobsters, and crabs. This means that sow bugs and pill bugs are more closely related to these animals than to insects like bees and ants. Crustaceans belong to a larger group of animals called arthropods, which also includes insects, spiders, and other animals.

Sow bugs and pill bugs breathe with gills, so they need moisture for respiration. But these tiny crustaceans have still had very successful lives on land, as there are now about 5,000 known species of pill bugs and sow bugs living just about anywhere from beaches to deserts. Wherever they go, these bugs eat decaying materials like plants and even dead animals; by following this diet, they help recycle nutrients in the soil.

Sow bugs and pill bugs, both of which are also called wood lice, look fairly similar to one another. A pill bug is commonly known as a "roly-poly" for its defensive behavior of rolling up into an armored ball. A sow bug, however, doesn't roll itself up and has a pair of tail-like structures on the end of its body.

MATERIALS

- Two clean, empty cardboard milk cartons
- Sharp utility knife
- Scissors
- Tape
- Dry soil
- Water
- At least 12 sow bugs or pill bugs (Details for collecting these are given in the activity below.)

PREPARATION

- Carefully use the utility knife and scissors to cut the tops of the two milk cartons completely off at least 4 inches (10 cm) up from their bottoms. Have an adult use the utility knife to cut into the cartons and then continue cutting them with the scissors. Discard (recycle) the top parts of the cartons. You should be left with two carton containers with open tops.

- About 1 inch (2.5 cm) up on each carton bottom, cut out rectangles that run the length of the side of the carton and that are between 1 to 2 inches (2.5 to 5 cm) tall. These will be the "doors" that the sow bugs or pill bugs will use to move between the cartons, so make sure these holes line up well between the two cartons.

- Turn the carton bottoms so that their rectangular door holes line up and then tape the two cartons together. Make sure there are no gaps through which any bugs could escape.

- Remove any large pieces of wood or rocks from the dry soil you'll be using. Put dry soil into one carton until it just reaches the bottom of the hole you made (or is about 1 inch [2.5 cm] deep). Smooth out the surface of the soil so it is flat. *Do you think the sow bugs or pill bugs will like this environment?*

- Put the same amount of dry soil into the other carton and then mix in a little water until the soil has a cookie dough consistency. Smooth out the surface so that it's flat, but do not press down hard because you don't want to compact the soil. *Do you think the sow bugs or pill bugs will like this environment more than the dry soil one?* Your habitat is now ready to do some experimenting.

PROCEDURE

- Go outside and collect at least a dozen sow bugs or pill bugs. (Only use one type of bug.) To find them, look under stones, decaying wood, old leaves, in gardens in the soil near plants, and along house foundations and basements. You may need to look closely because they're small, can blend in well with their surroundings, and might freeze in place when startled, making them harder to spot.

- Put an equal number of the little crustaceans in each carton (meaning there should be at least six in each). Allow them to explore the cartons while you observe for at least 30 minutes. *How many bugs are in each carton over time? Does the carton that most of the bugs are in change over time? What other types of behaviors do you observe?*

- *Overall, which environment do the sow bugs or pill bugs seem to prefer, the one with damp or dry soil?*

SCIENCE FAIR IDEA

Repeat this activity but compare different environments, such as leaf litter versus soil, or small rocks versus small pieces of wood, etcetera. *Which environment do the sow bugs or pill bugs prefer in other environmental pairings?*

EXTRA

You could try to quantify your results from this activity. To do this, every five minutes you could count how many sow bugs or pill bugs are in each habitat. *How do the numbers of these crustaceans in each environment change over time?*

EXTRA

You could do similar activities with other small, common animals, such as crickets, earwigs, ants, slugs, snails, mealworms, and wax worms. (Crickets, mealworms, and wax worms can usually be purchased at a local pet store, whereas the other animals may be found locally by looking under rocks, garden plants, rotting logs, and leaf litter.) *What type(s) of environments do these other animals prefer?* (If the animal you're using might be able to escape from the containers, be sure to secure a lid to them!)

OBSERVATIONS AND RESULTS

Did most of the sow bugs or pill bugs end up in the damp soil environment?

Sow bugs and pill bugs are crustaceans, just like shrimp, lobsters, and crabs. They breathe with gills, so they need moisture in order to breathe. Because of this, you should have seen that most of the sow bugs and pill bugs spent more time in the damp rather than the dry soil environment. After about 30 minutes, you may have even seen several of the little crustaceans actually dig down into the damp soil and settle in there. The sow bugs or pill bugs that spent time in the container with the dry soil probably only stayed there temporarily before going back to the damp soil. Although sow bugs and pill bugs are amazing in that they're crustaceans that have adapted to live on land, they still need moisture to breathe and survive.

CLEANUP

When you are done observing the sow bugs or pill bugs, you may release them where you found them. Be sure to rinse and recycle the empty milk cartons.

Snuggly Science
How Puppies Keep Warm

CHECK OUT THIS ADORABLE TEMPERATURE TEST!

Have you ever woken up on a cold, frosty morning and just wanted to snuggle down deeper under the covers? Other mammals, such as puppies and piglets, do not like being cold either, but they do not have hands or blankets to wrap themselves up. So when these animals get chilled, they change their behavior and do things like huddle—curl up close to other animals. In this science activity, you'll see just how much huddling can help reduce heat loss.

PROJECT TIME

45 minutes

KEY CONCEPTS

Heat
Temperature
Physics
Animals

BACKGROUND

What makes puppies want to sleep in an adorable heap? Snuggling, also known as huddling, does more than create cute puppy piles. Puppies, like other animals, huddle to keep warm. Heat is continuously flowing into and out of everything around you. Just as a ball rolls downhill, heat rolls away too, flowing from an object that's hotter to one that's colder. If, for instance, you put a cup of hot chocolate in a room-temperature room, the heat will flow from the hot chocolate to the room, cup, and table. These surroundings will actually warm up (ever so slightly) from the hot chocolate's heat. When the temperature of the hot chocolate cools enough to reach the surrounding temperature, the heat flow stops.

Heat flows from a hotter object to a colder object in several different ways. For example, when puppies are huddling together to keep warm, they are reducing how much of their bodies (specifically surface area) are exposed to the colder open air, and so are reducing heat loss in two ways: via radiation and convection. A puppy also gains heat another way—conduction—if the puppy it's snuggled up with happens to be warmer.

MATERIALS

- Four glass jars or bottles that are the same size and shape, with lids that can be discarded (The lids will be punctured in this activity. Use jars whose lids do not stick out farther than the glass sides of the jars. If using canning jars, you will need rings for the lids.)
- Thermometer with a stem that has an upper range of at least 200°F (93°C), such as a candy thermometer
- Large soup pot
- Nail
- Hammer
- Measuring cups
- Water
- Oven mitts
- Towel
- Timer or stopwatch
- Piece of paper and pen or pencil

PREPARATION

- Using a hammer and a nail, carefully make a hole in the center of each jar's lid just big enough for the thermometer's stem to fit through. If needed, slightly adjust the shape of each hole (using the nail) so that the thermometer's stem sticks straight—vertically—down into the jar. *Why do you think you'd want to measure the core temperature of the jars?*

- Fill the jars about three-fourths full with equal amounts of water (of the same temperature for each jar). Each jar will be a model of a puppy.

- Fill the large soup pot with enough water so that when three jars are in the pot, the water level within the jars equals the water level outside of the jars (in the pot).

- Be careful when working with the hot water. Adult supervision is required when using the stove.

PROCEDURE

- Set a timer for 10 minutes or have a stopwatch ready.

- Put the large soup pot on the stove and put one jar inside. Put the thermometer inside the jar (going through the lid's hole) so that the thermometer is held upright inside the water in the jar. Turn on the stove.

- When the temperature of the water inside the jar reaches 190°F (88°C), turn off the stove, immediately remove the jar from the water bath using oven mitts and dry the outside of the jar quickly on a towel.

- Immediately start the timer or stopwatch. Write down the water temperature inside the jar when you do this. Then continue taking water temperature measurements every two minutes for at least 10 minutes. *How does the temperature change over time?*

- Repeat these steps, but this time put three jars inside the pot (instead of just one). Put the thermometer into one of the three jars. When the jar reaches 190°F (88°C), remove all of the jars, quickly dry them and arrange them so that two sides of each jar are touching two other jars, forming a triangle. Write down the initial temperature in the jar, and then again take temperature measurements every two minutes for at least 10 minutes. *How does the temperature of the jar huddled with two other jars compare with that of the one jar that stood alone over time? At what time was the difference in their temperatures the greatest?*
- *Overall, what do your results tell you about how and why puppies and other warm-blooded animals huddle to keep warm?*

EXTRA

Try repeating this activity but compare three jars touching with two jars touching or with three jars separated by 1 inch (2.5 cm). *How does the heat loss compare between these different configurations as well as with the solitary jar?*

EXTRA

Think of how you could add "fur" or "fat" to your puppy models in this activity and then try it out. *How do these additions affect heat loss?*

EXTRA

You could investigate increased heat loss through convection by blowing a fan on the jars. *How does this affect how much heat is lost over time?*

OBSERVATIONS AND RESULTS

Did the three jars huddled together lose less heat over time compared with the one jar alone? Was their difference in temperature the greatest at the end of the 10-minute period?

Over time, you should have seen that the water in the jars—alone and in a huddled group of three—lost heat, but the single jar lost heat faster than the three jars huddled together. The temperature of the water in the jars should have started out at about 190°F (88°C), because this was the temperature immediately when they were removed from the pot. As just one example, the temperature of the water in the solitary jar may have then dropped to about 172°F (78°C) after 10 minutes whereas the temperature of the water in a jar huddled in a group of three may have only dropped to about 176°F (80°C). (The exact size and shape of the jars affects how quickly they lose heat, as a larger amount of hot water will take a longer amount of time to cool down compared with a smaller amount of the same-temperature water; and the shape of the jars affects their total surface area.) When the jars are close together, just like puppies snuggling together, they help keep each other warm, primarily through conduction.

CLEANUP

Dry your jars to use again. Return all other materials to where you found them.

Bird-Spotting Science

Predict a Bird's Lifestyle Based on Its Feet

STUDY ANIMAL ADAPTATIONS WHILE YOU WATCH FOR BIRDS IN YOUR AREA!

Have you ever wondered how different animal species can survive—and thrive—in all sorts of environments, even when the animals seem closely related? For example, polar bears swim in the Arctic Ocean where they face negative temperatures, whereas other bears live in warm southern California. How do they do it? The answer is adaptations! Their bodies have special features that allow them to live in their environments. In this activity you'll investigate what the adaptations of birds in your area tell you about those birds' lifestyles. Get ready to do some bird-watching!

PROJECT TIME

1 to 2 hours

KEY CONCEPTS

Animals
Biology
Adaptations
Evolution

BACKGROUND

If you were an animal with a thick layer of fat under your skin and a heavy coat of fur, such as a polar bear, would you live in a tropical forest? No way! You'd be way too hot! But if you lived in the cold of the Arctic, near the North Pole, these adaptations would be necessary to keep you warm. All creatures have adaptations. Some adaptations help animals deal with the climate, such as the polar bear example above. Other adaptations help animals move better in their environments, blend in with their surroundings so they don't get eaten by predators, or successfully hunt for their dinner.

Birds are animals with a lot of specific and useful adaptations. A major one is the shapes of their feet. Have you ever noticed that one kind of bird's foot can be wildly different from another type's? For example, chickens and ducks are both birds, but they have very differently shaped feet. Chicken feet have relatively long skinny toes with strong, sharp nails, whereas ducks have webbed feet. Chickens' feet allow them to scratch at the ground to find insects and seeds to eat, whereas ducks, who live in the water, have webbing between their toes that helps them paddle.

MATERIALS

- Unless you're going to a zoo or pet store soon, you'll want a bird-watcher's field guide to identify birds in your area. Alternatively, you could use online resources to identify the birds you see.
- A place to observe birds. (This could be your backyard—especially if you have a bird feeder or birdbath—or a park, field, forest, beach, aviary at a zoo, pet store, or farm that has different types of birds.)
- Binoculars, which are only needed if you will be watching birds from a distance. Inexpensive ones are fine.
- Sheet of paper
- Pen or pencil

PREPARATION

- Get ready to do some bird-watching. Unless you're going to a zoo or pet store, have a local bird field guide ready or have access to an online resource so you can identify local species.

PROCEDURE

- Bring a sheet of paper, a pen or pencil, and your field guide (or access to an online resource) to one of the bird-observation places you decided on. If needed, also bring binoculars. Observe the birds there. (Be sure to have an adult accompany you.) *What kind of birds do you see?*

- Identify the different species of birds you see. (If you are at a zoo, pet store, or another place where the bird species are labeled or simple to identify, this will be easy.) Use a field guide or online resource to do this.

- *What species do you see?* On your sheet of paper, write their names down.

- Look at the feet of each bird species. *What type of feet do they have?* Write down one of the following seven types of feet for each species:
    - Climbing: Two toes in front, two toes in back
    - Swimming: Webbed feet
    - Running: Strong-legged with two or three thick toes, all facing forward
    - Perching: Three toes in front, one toe in back
    - Grasping: Clawlike feet with curved talons
    - Scratching: Four toes, all with strong nails for digging into the ground
    - Wading: Long, thin legs and toes

- **Tip:** If it is difficult to clearly see any bird's feet, you could look for pictures (online or in your field guide) of birds that more clearly show their feet.

- Try to make observations for at least five species. The more you observe, the better your results should be!
- Based only on the type of feet each bird has, predict what kind of lifestyle you think that bird has. *In other words, where do you think it spends most of its time? Is it usually perched in a tree, climbing trees, walking on the ground, or swimming in water?*
- Do some research on each species of bird to find out where it actually spends its time. To do this, you could use the field guide or online resources. You'll want to look into where each bird nests, roosts (sleeps), and forages (gets food). *Where does each bird spend most of its time?*
- *Based on your results, how accurate an indicator was each bird's foot type for the kind of lifestyle it usually leads?*

EXTRA

In this activity, you looked at the general lifestyle of birds, but there are many aspects that go into birds' lifestyles, such as where they nest, where they roost, and how they get food. Pick one of these aspects and investigate again how well a bird's foot type is as an indicator of that part of its lifestyle. *How good is a bird's foot type at, for example, predicting how it gets its food?*

EXTRA

Try this activity again, but this time compare the birds' beaks instead of their feet. *How well can a bird's beak be used to predict its lifestyle and/or specifically what it eats? Do birds with different types of beaks eat different things?*

EXTRA

Pick two diverse habitats, like the beach and a forest, and investigate the species of birds in those areas. *Do birds in the same habitat have similar types of feet? How about birds in different habitats?*

OBSERVATIONS AND RESULTS

Were you able to use the type of foot that each species of bird has to accurately predict that bird's lifestyle?

Chickens' feet are adapted to allow them to scratch at the ground and find insects and seeds to eat, whereas ducks' feet are adapted to help them paddle in the water. In this activity, you should have seen that a bird's feet are usually a good indicator of the type of lifestyle it leads. For example, the common house sparrow (*Passer domesticus*) has perching-type feet, and it indeed spends a lot of time perching on trees, other plants, and structures in different environments. Similarly, lovebirds (of the genus *Agapornis*), African grays (*Psittacus erithacus*), cockatiels (*Nymphicus hollandicus*), and other parrots you might find in pet stores and aviaries have climbing feet. Not only do they frequently climb in captivity, but in the wild they also spend much time doing this—often climbing on and nesting within trees in wooded areas. The type of feet a bird has, however, might not reflect every aspect of its lifestyle. For example, many birds have climbing- or perching-type feet but mostly forage on the ground. This includes mourning doves (*Zenaida macroura*), the American robin (*Turdus migratorius*), and some kinds of parrots, along with many other bird species. These adaptations might still be left over from their ancestral species that lived in different environments.

CLEANUP

Put any materials you used back where you found them.

Species Science

Exploring Bug Biodiversity

HOW MANY DIFFERENT SPECIES OF BUGS CAN YOU FIND IN AN AREA?

It might seem impossible to estimate how many different bugs live in a nearby field—or how many types of birds live in a forest. But wildlife biologists often have to try to determine the number of different species in a given habitat. It might sound challenging, but it is important. This can help us understand how healthy the habitat is. In this activity, you'll get to take on the role of a wildlife biologist by examining the biodiversity of bugs and other small invertebrates (such as spiders, centipedes, and roly-polies) in your neighborhood or backyard using a homemade bug vacuum!

PROJECT TIME

1 to 2 hours

KEY CONCEPTS

Animals
Environment
Biodiversity
Ecology
Habitat

BACKGROUND

Ecology is the study of living things and the habitats they inhabit. In ecology, the more different species that live in a given habitat, the healthier it is considered to be. The degree of species variety found in a habitat is referred to as its biodiversity. In a given habitat, the plants and animals normally rely on one another for their entire life cycles. For example, let's say berries from a bush are eaten by a field mouse, then the mouse helps transport the seeds inside the berries to different locations, thereby allowing more berry bushes to grow. The mouse might then be eaten by a bird of prey, like a hawk, and on and on the chain goes. This web of relationships is called an ecosystem. The greater the biodiversity, the larger the ecosystem, and the healthier the habitat is thought to be.

Wildlife biologists study and monitor the health of habitats and ecosystems. One way they do this is to survey (identify and count) populations of plants and animals to determine how much biodiversity is in a habitat. In this activity, you'll do your own local biodiversity survey by making a "bug vacuum" to collect as many species of bugs as you can find.

MATERIALS

- Plastic container with lid, approximately 1 cup or 1/2 pint size (about 237 ml)
- Single-hole punch or leather awl (Use caution and have an adult help with sharp objects.)
- Two flexible drinking straws (Wider ones are better as they will allow larger bugs to be caught.)
- Tape
- Pen or pencil
- Nylon stocking that can be cut up
- Scissors
- Clear plastic wrap
- Access to an area in your yard, a field, or another outdoor place where you think bugs might live. **Tip**: These types of small invertebrates can often be found under rotting wood, stones, and decaying leaves. And if it is hot and dry outside, more bugs may be found in shady and/or moist areas.

PREPARATION

- With the help of an adult, make a hole in the side of the plastic container (using a single-hole punch or carefully using a leather awl). The hole should be about 1/2 inch (1.3 cm) below the top rim of the container. Make sure that when the container has its lid on, it does not block the hole.

- Using the same technique, make a hole opposite the first one, again about 1/2 inch (1.3 cm) below the container's top rim.

- Insert the mouth end of a flexible straw into one of the holes. Stick the straw about 1 inch (2.5 cm) into the container. If the straw does not fit tightly into the hole, wrap tape around the straw to make it fatter until it fits snugly into the hole. If the straw is too large for the hole, gently push a pen or pencil through the hole to widen it until the straw just fits. *Make sure the connection between the straw and the hole is snug. Why do you think this is important?*

- Cover the mouth end of the second straw with a small piece of nylon stocking. *Why do you think it's important to cover one straw like this?* Use tape to secure the stocking piece to the straw. If necessary, overlap two layers of the stocking so that air can pass through, but bugs can't.

- Insert the covered end of the second straw into the second hole in the plastic container. Again, make sure the fit is snug, and stick the straw about 1 inch (2.5 cm) into the container.

- Cut out the center of the plastic container's lid. You can do this by folding the lid in half and cutting out the center so that only a 1/2-inch (1.3 cm) border around the lid's rim remains. This will be the observation window of your bug vacuum.

- Stretch a piece of clear plastic wrap over the top of the plastic container. Then snap the container's cut-out lid back on to hold the plastic wrap in place. Make sure there are no holes in the plastic wrap or your subjects could escape!

PROCEDURE

- Bring your bug vacuum out to an area in your yard, a field, or another outdoor place where you think bugs might live. *What types of bugs do you think you'll find here?*

- If you want, you can use sticks and/or rocks to mark off a small area that is roughly 4 feet by 4 feet (1.2 m by 1.2 m). This will help you define the area you are surveying for bug biodiversity.

- Carefully search a small area for bugs (such as the one you may have marked off). You will need to have a search pattern so that you do not crush the resident bugs before you suck them up, and also so you know what parts you have checked already. (For example, you could start at one corner of your area and walk along its edge, looking into the area you're surveying, and then carefully check the area inside, only walking on places you've already collected bugs from. Alternatively, you could break your area up into smaller sections and check each one at a time.) *Do you see many bugs?*

- When you find a bug, take your bug vacuum and gently place the filterless straw so that the end is very close to the bug. Place the other straw with the nylon filter in your mouth and breathe in forcefully. This will, like a household vacuum, suck in the bug. Repeat this process until the bug has been transferred into the plastic container chamber.

- Repeat this process until most of the bugs in your small surveying area have been collected. **Tip:** Don't forget to check for bugs under wood, rocks, and leaves—but get ready for them to quickly scatter!

- Examine the bugs you collected by looking at them through the plastic wrap window of the bug vacuum. *How many different types of bugs did you collect? How many of each type did you collect? Can you identify the different types of bugs you caught? Based on your data, do you think there is a lot of biodiversity in the area you chose to survey?*

EXTRA

Research which species of bugs live in your area, then repeat this activity using bug guidebooks and websites to identify the samples you collected. *Did the kinds of bugs you expected to find in your backyard match your results?*

SCIENCE FAIR IDEA

Compare two or more different habitats, such as a field and a forest. *Do different locations have different bugs? Which habitat has more bug biodiversity? Why do you think this is?*

EXTRA

Try changing the conditions in your survey area, such as by sprinkling the ground with water before doing your survey or waiting until just after it has lightly rained. *Does changing conditions like this affect the biodiversity you find?*

OBSERVATIONS AND RESULTS

Did you find many different types of bugs in the area you surveyed? Were there many ants, beetles, earwigs, and/or other bugs—or things that looked like bugs?

If you surveyed an area that has good biodiversity, you should have found many different types of bugs. If you did not, the area may have been too small or conditions may not have been right for supporting a diverse population of bugs. Some of the most common types of bugs in North America include various flies, beetles (including ladybugs), butterflies and moths, ants, bees, wasps, centipedes, millipedes, earwigs, grasshoppers, and crickets, among many others. You may have also found spiders, pill bugs (also called roly-polies), wood lice, centipedes, and/or millipedes—all of these are not technically insects but, like bugs, they are also small invertebrates. Spiders are arachnids, and pill bugs and wood lice are a type of crustacean called isopods. Depending on the exact type of area you surveyed, you may have found different types of bugs, because different species prefer various kinds of conditions.

CLEANUP

Release the bugs back outdoors, preferably close to the area in which you found them.

Scoping Science

How Do Animals Look for Food?

USE SEEK-AND-FIND BOOKS TO TEST A SCIENTIFIC THEORY ABOUT HOW ANIMALS HUNT!

PROJECT TIME

30 minutes

KEY CONCEPTS

Foraging
Behavior
Visual search
Search image

You likely have food preferences—perhaps you love salty snacks or dislike red meat—but in general, humans eat many different foods. For that reason, our species is considered omnivorous. But some animals are much more choosy. Herbivorous rabbits, for example, dine only on plants, whereas carnivores, such as sharks, feast principally on animal prey. Locating these very different types of foods can obviously require different tactics. In this activity, you'll learn about a theory that biologists use to explain different strategies for foraging, or finding food. You'll also apply this theory to test your visual hunting skills.

BACKGROUND

Believe it or not, animals don't just pursue the first food option they see. Biologists who study animal behavior have found that different species have different strategies. This is because locating a meal isn't always easy—it requires time and energy. To maximize the amount of energy an animal gets from its prey, it needs to pursue food nutritious enough to make up for the time and energy it needs to catch and consume a meal.

Scientists call this "optimal foraging theory"—or the best strategy for finding a meal—and they have a simple mathematical formula to express it: E/h + s. "E" stands for energy, specifically the energy an animal gains from eating food. The "h" stands for handling time—this represents how long it takes to get the food ready to eat. An apple, for example, has very little handling time, but an oyster requires more effort. The "s" is for search time, which can vary depending whether food is rare or plentiful, or when food is hard to detect or sticks out easily. Strategies shift based on an animal's abilities and environment. In this activity, you'll measure search times and experience one aspect of optimal foraging theory firsthand.

MATERIALS

- *Where's Waldo?* or other visual search book (you can find these books at your local library)
- Pencils and paper
- A partner (groups of two or three individuals work)
- A clock or stopwatch

PREPARATION

- This activity works best with a visual search book such as *Where's Waldo?* in which you repeatedly seek out a specific target.
- To make the activity more challenging, do not peek at the puzzles before it is your turn to "hunt."

- Ask a partner to be the first hunter. Ask them to not peek at the image search until you say, "Go!"
- Tell the "hunter" what target or "prey" they will seek out in multiple puzzles (Waldo, for example, is a good target).

PROCEDURE

- Record the "hunter's" name and their target prey.
- Hit the stopwatch or start watching the clock when you say, "Go!" Time how long it takes your partner to find their prey and record the total time needed. *Are you surprised by how quickly or slowly your partner found the target?*
- Switch roles with your partner and repeat the previous steps. *Was finding your target easier or harder than you expected? What made the task challenging?*
- Switch back and forth until you have each "hunted" your prey at least five times. Every time it is your turn, try a different puzzle page but seek out the same target. *Does changing environments make the puzzles easier or more difficult? Do you think you get better at finding the same target in different environments?*
- So far, you have been hunting as a specialist—you have just one kind of prey. Start the activity again and repeat as a generalist, meaning you can hunt different kinds of prey. In this version, pick a different target every time you hunt, but once again use a new puzzle page each time. Repeat five times each and record your times. *Do you notice any differences as a generalist than a specialist?*

EXTRA

If you want to compare one more variation, try hunting for a totally different target in puzzle pages you have already seen. *Is it easier to hunt in familiar territory?* Next, look at your times as both a generalist and as a specialist. *Did either you or your partner's hunting skills improve? Did times increase or decrease as you repeated the search?* Foraging theory suggests you need to select foods that give you enough energy to make up for that you lose in the time spent hunting and catching food. *Which prey items took the most time to find? Following the theory's logic, what would make hunting these items worthwhile?*

EXTRA

Because you've mostly played this activity with time in mind, create your own variation that tests another component of optimal foraging theory: energy. With your partner, decide on different energy values for different targets, then revisit old puzzle pages and give each other 10 minutes to collect as many prey items as you can find. In this version, each hunter needs to declare what they are searching for and can only collect points by finding a mentioned target. *How do you and your partner match up? Are there strategies that work in your favor, such as seeking out targets with a certain value?*

OBSERVATIONS AND RESULTS

Did seeking the same target as a specialist get easier over time compared with seeking new targets in each puzzle?

In this activity, your visual search was very similar to that used by a hunting hawk as it soars high above the ground. As you probably noticed, trying to identify your target in such a cluttered landscape was not easy. Yet you may have also found that by searching for the same target repeatedly, you became more adept at looking for that particular kind of prey. This is a common result and something that scientists have discovered in birds of prey. Biologists call this trick developing a "search image," because it allows the animal to have a very specific visual image in mind when scanning a landscape. Changing your target every time, however, is more taxing on your brain and makes a long search even longer.

CLEANUP

Put any materials you used back where you found them.

Signaling Science

What Household Solutions Repel Ants?

DON'T GET ANTSY! TRY TO KEEP ANTS AWAY FROM YOUR PICNIC WITH DIFFERENT SOLUTIONS AND PRODUCTS.

Have you ever had ants ruin your picnic? Commercial ant repellents can keep them away, but who wants to spray poison near their food? In this activity, you can investigate the effectiveness of some less toxic solutions that you may have around your home. Armed with your discoveries, you may be able to keep your next picnic from turning into an ant buffet!

PROJECT TIME

1 hour

KEY CONCEPTS

Biology
Chemistry
Chemical signals
Entomology
Repellents

BACKGROUND

All animals sense chemical signals in their environments. Some animals use these signals to communicate. One animal famous for chemical communication is the ant. When you see a line of ants, it is because they are following an invisible chemical trail. This type of chemical signal is positive—it is called an attractant because it is meant to attract others. But signals can also be negative, warning other ants to stay away. This type of signal is called a repellent.

Sometimes repellents can come in handy. We don't like ants to get into our homes, so if we apply a negative chemical signal around the foundations, the signal will tell the ants to keep out. Chemical companies will sometimes try to copy a negative signal to sell as an insect repellent. But sometimes these chemicals can have side effects or be poisonous to pets or small children. Luckily, there may be less toxic remedies around your home.

MATERIALS

- Cotton swabs
- Small cups or bottle caps
- Large, vinyl tablecloth
- A location outside where there are plenty of ants crawling around
- Water
- Sugar
- Baking soda
- Vinegar
- Detergent (dish soap, laundry detergent, hand soap, etcetera)
- Hot pepper sauce
- Lemon juice (or another citrus juice)
- Commercial ant repellent, if you want to compare its effectiveness
- Any other solution you would like to test!
- Watch or stopwatch (optional)

PREPARATION

- In a small cup or bottle cap, make a saturated solution by dissolving some baking soda into water. Keep adding and mixing baking soda until no more will dissolve.
- In a small cup or bottle cap, make sugar water by dissolving sugar in water until no more sugar will dissolve.
- If you use a commercial ant repellent, you should have adult permission and supervision.
- Separately make up additional small cups or bottle caps of the solutions you will be testing: Water, vinegar, detergent, hot pepper sauce, lemon juice, and, if you choose to test it, commercial ant repellent.

PROCEDURE

- Lay out your vinyl tablecloth on the ground in an area of heavy ant activity. Wait awhile for the ants to walk on the tablecloth. (Be careful when working with ants—some can sting and bite!)
- **Tip:** If not many ants are walking on the tablecloth, you could try putting a small piece of fruit near them, such as an apple slice. Wait until the ants cover it, and then carefully shake them off onto the tablecloth. If it is very hot outside, you may need to wait until later in the day when it cools down for ants to come out of their holes.
- Next to the tablecloth, have ready the solutions you will test: saturated baking soda solution, sugar water, water, vinegar, detergent, hot pepper sauce, lemon juice, and, if you want, chemical ant repellent.
- Test a solution on an ant by dipping a cotton swab into the solution and drawing a circle around the insect, about 1 inch (2.5 cm) in diameter. *Did the ant stay inside the circle? Was there anything else you observed about the ant's behavior?*

- In this way, test solutions on the ants until you have tested at least five ants with each substance.
- *Which solutions usually kept ants inside the circle? From which did the ants readily walk out?* Potential repellents will be solutions that clearly trap the ants inside a circle.
- *Were there circles in which the ants spent time standing, neither quickly leaving nor appearing trapped?*

SCIENCE FAIR IDEA

Once you find some potential repellents, you can figure out which ones work best. Try this activity again, but now time how long the ant remains trapped in the circle. *Which repellents repel the ants for the longest amount of time?*

SCIENCE FAIR IDEA

You can also time how long the ant stays trapped in the circle to experiment with different concentrations of repellents. *How diluted can your solutions be and still act as effective ant repellents?*

EXTRA

Try a similar activity to this one but only test dry or powdered substances. You won't need a tablecloth for this. For a hard, dry substance like chalk, you can draw the circle on a sidewalk. For a powder, like baking soda, sugar, or salt, just sprinkle it in a circle around the ant. *Can you find some powdered or solid substances that work as ant repellents?*

OBSERVATIONS AND RESULTS

Did some solutions keep the ants trapped in the circle, whereas the ants quickly walked over circles made of other solutions? Did the baking soda, detergent, and hot pepper sauce clearly repel the ants, although the water did not?

Ants rely on chemical signals to navigate toward food, their nest, and other places, and we can use negative signals, or repellents, to discourage ants from going somewhere. Baking soda, vinegar, lemon juice, some detergents (and cleaning products), and hot pepper sauce (and other spicy substances, such as red chili pepper, black pepper, and cayenne pepper) usually repel ants to varying degrees, and you may have seen them trapped in these circles. Many other common household solutions can also repel ants, including cinnamon, mint, salt, cloves, garlic, onions, and bay leaves. Many ants are attracted to sugar, so you may have seen ants stop and spend some time on the circle made up of the sugar water—they may have been enjoying a snack! The ants should have quickly crossed the circles made of water, but if thick lines were made using the water or other solutions—or if the ants were thirsty—this could cause them to hesitate before exiting the circles.

CLEANUP

When you are done with this activity, let the ants return to where you found them. Wash off the vinyl tablecloth and be careful and thorough if you chose to use commercial ant repellent.

Squirmy Science

Which Soil Types Do Earthworms Like Best?

DISCOVER MORE ABOUT THE FASCINATING EARTHWORM AND WHERE IT LIKES TO MOVE AND EAT!

Have you ever dug a hole in the soil and noticed earthworms wriggling out of your way? Not only are earthworms good for birds and fish to eat, but these little animals actually work hard to help put food on your table too. By adding nutrients to the soil they live in, earthworms are continually working to keep soil healthy so that most plants can grow well in it. To do their important work, earthworms prefer certain types of soil over others.

PROJECT TIME

30 minutes of preparation, plus 1 day

KEY CONCEPTS

Biology
Zoology
Nutrients
Plants

BACKGROUND

Here's a riddle for you: What has five hearts, no eyes, and helps feed all the people on Earth? The earthworm! This amazing animal is important for keeping soil rich in nutrients. Just as you need vitamins to stay healthy, such as the vitamin C you get from fruits and vegetables, plants need nutrients to grow properly as well—and they get them from the soil. The earthworm is able to transform dead plants and leaves (also called plant matter) into nutrients for the soil. Earthworms can make food for living plants from dead ones. They leave this food (the nutrients) behind in the soil where growing plants can absorb it.

In less than a day, an earthworm can eat its entire weight in dead plant matter. So, for example, if you were an earthworm and weighed 50 pounds (22 kg), you'd need to eat about 50 pounds (22 kg) of food every day! During this moving feast, an earthworm is also loosening and aerating (letting air get into) the soil to make it easier for many beneficial plants to grow in it.

MATERIALS

- A cool area that is not brightly lit
- Measuring cups
- Dry sand (1 1/2 cups)
- Dry gravel (1 1/2 cups)
- Dead leaves, crumbled (1 1/2 cups)
- Dry potting soil (1 1/2 cups)
- Mixing bowl
- Spray bottle
- Water
- Spoon or fork
- Empty plastic box with a sealable lid, larger than a shoe box, but smaller than a laundry basket
- Earthworms (at least five); available at a bait shop, or use a small hand trowel or shovel to dig up earthworms from the ground (usually in shady, moist soil), or find them on the ground after it has rained
- A small container with a lid in which to hold the earthworms; put some damp soil and dead leaves inside the container for the worms, and keep them in a cool place until you are ready to do the activity

PREPARATION

- You can test this activity indoors or outside. Choose a location that is cool and in the shade (if you are testing outside) or in a dimly lit room (if you are testing indoors). If the plastic box is transparent, use an especially dark location, such as a closet.
- Put 1 1/2 cups of sand into the mixing bowl. Spray the sand with water (counting the squirts as you go) until it all feels wet, stopping before puddles of water appear. Mix the moistened sand with a spoon or fork.
- Transfer the wet sand from the mixing bowl to one corner of the plastic box. Make sure the sand does not spill out over one-quarter of the area of the bottom of the box.
- Rinse out and dry the mixing bowl. Spray the other "soils" (gravel, dried leaves, and potting soil) in the mixing bowl (each separately) just as you did the sand. Use the same number of sprays of water for each soil type as you did for the sand. When each soil has been sprayed, transfer it in its own corner of the box. *How wet does the sand feel compared with the gravel, dried leaves, and potting soil?*
- As you fill the four corners of the plastic box with the different "soils," make sure that none spill over and touch the other soils. Space the soils so that there is roughly 1 to 2 inches (2.5 to 5 cm) of space between the samples.

PROCEDURE

- Place your plastic box in the cool, unlit area you chose. Check to make sure the soils are still separated and not touching one another.
- Place your five (or more) earthworms in the center of the box, in the middle of the four soil samples. Watch the earthworms for a minute or two. *What do the earthworms do at first? How do they move? Do they follow one another? Did they all go into one soil or did they burrow into different soils?*

- After watching them for a minute or two, close the lid on the box. Note what time it is. Let the earthworms stay undisturbed in the box for 24 hours.
- After 24 hours have passed, open the box. *Do you see any worms outside the soil?*
- Carefully go through each pile of soil to look for the earthworms, moving the soil to another container or an empty part of the box. You may need to look very closely through the different soils to make sure you do not miss a small earthworm. *Which soils have the most earthworms in them? Which soils have the least? Why do you think some of the soils drew more earthworms than others?*

SCIENCE FAIR IDEA

Do earthworms stay in the soil in which they are put? Do this activity again but put one earthworm in each soil type. *After 24 hours, is there still one earthworm in each sample, or are there more earthworms in one soil compared with another?*

EXTRA

Earthworms are blind, but they have special cells that can sense light. *What kinds of light do earthworms avoid and what kinds do they not mind?* Set up a box with damp potting soil in each corner. Cut out small holes in the box's lid at three corners. Put a small red-light flashlight pointing straight down above one corner, a small blue-light flashlight above another corner, and a small white-light flashlight above the third corner. Put an earthworm in the center of the box, put the lid on, and turn on all three flashlights. *Which flashlight (or flashlights) does the earthworm go toward and which does it avoid? Did the earthworm try to stay away from all the lights or just certain colors? Do multiple earthworms behave the same way?*

OBSERVATIONS AND RESULTS

Were most of the earthworms in the leaves? Were some in the sand and potting soil? Were none in the gravel?

Earthworms eat dead plant matter, such as fallen leaves, and transform it into a nutritious substance that plants can absorb more readily. So, if they want a tasty meal, they would most likely be in the leaves. However, if there is organic matter (dead plant matter or animal matter, which they also eat) in the other soil types tested, the earthworms might be attracted to those soils as well. But just because an earthworm is in one of the soils does not mean that it is its ideal environment or that it wants to eat that soil. For example, you may find one or more earthworms in the gravel, but this could be due to factors other than the earthworm searching for a food source—the gravel's humidity, for example, might be more appealing or there may be possible chemoattractants on the gravel.

Besides adding nutrients to the soil, earthworms also keep the soil healthy by tunneling. As they eat, earthworms create tunnels that keep the soil open and full of paths for water and air. Without earthworms, the soil would be smashed down, or compacted, making it difficult for water and air to reach the roots of plants. Overall, it would be very hard to grow all the beautiful plants that give us oxygen and food without these wiggly little creatures.

CLEANUP

When you're done with this activity, gently and carefully return the earthworms to some soil outside.

Porous Science

How Does a Developing Chick Breathe Inside Its Eggshell?

TRY AN EGG-CELLENT ACTIVITY AND FIND OUT MORE ABOUT CHICKEN EGGSHELLS!

PROJECT TIME

1 hour, 20 minutes

KEY CONCEPTS

Biology
Respiration
Porosity
Food science

Have you ever wondered how an unborn chick breathes inside its shell? Every animal needs oxygen to live, so the chick must get air somehow! When an animal—including a human—inhales, oxygen enters its lungs and is then distributed to all the different parts of its body. The animal's metabolism converts the oxygen into energy. During this process, a waste gas called carbon dioxide is produced. To get rid of it, the carbon dioxide is carried back to the lungs, where it is collected and exhaled. So not only must the chick have a way to let oxygen in, it also must somehow let carbon dioxide out. How does it do this sealed inside an eggshell?

BACKGROUND

When oxygen enters an animal's lungs, it is distributed by the bloodstream. It is also the bloodstream that carries carbon dioxide back to the lungs to be breathed out. Animals that grow inside their mother, like humans, get their oxygen directly from their mother. The bloodstream of the baby animal and the mother are connected via an umbilical cord, which allows the baby to collect oxygen that their mother breathes in as well as use the mother's lungs to expel the carbon dioxide.

How do animals such as chickens, which develop inside an egg outside of their mother's body and therefore do not have umbilical cords, take in oxygen and get rid of carbon dioxide? Bird eggs and some reptile eggs have a hard shell. Directly under the shell are two membranes. Between the membranes is a small air cell, also called an air sack, filled with oxygen. As the animal develops, it uses the oxygen, which must be replenished, and it also has to release carbon dioxide. How does this happen? Well, if you examine a chicken egg carefully with a magnifying glass, you'll see that there are tiny little holes, called pores, in the shell. In this activity, we'll see how those work to let the developing chick breathe.

MATERIALS

- Large pot or bowl
- Water
- Blue food coloring
- Liquid dishwasher detergent
- Teaspoon measurers
- Three eggs (for best results, do not use freshly laid eggs; rather, use older, commercial eggs)
- Tongs or large spoon
- Cup
- Plate or paper towel
- Optional: a sensitive scale, such as a digital kitchen scale or a triple-beam balance that can measure tenths of a gram

PREPARATION

- Pour 1 1/2 cups (355 ml) of water in a large pot or bowl.
- Add 1/4 teaspoon (1.2 ml) of liquid dish detergent and 1/4 teaspoon (1.2 ml) of blue food coloring. Mix well.

PROCEDURE

- Carefully put the three eggs in the pot with the water, dish detergent, and blue food coloring.
- Make sure that the eggs are submerged in the liquid. If part of the egg is above the surface of the water, mix together liquid dish detergent and blue food color with more water in the same proportions as you did before. Add this to the pot until the eggs are submerged.
- Set a timer for one hour or make a note of the time.
- After the eggs have soaked in the liquid for at least one hour, carefully lift one of them out of the liquid using the tongs or large spoon. *How does the egg look?*
- Crack the raw egg into a cup, being careful not to damage or crush the shell much.
- Set the empty eggshell on a plate or paper towel.
- Carefully inspect the inside of the shell. *What do you see?*
- Crack open the other two eggs in the same way. Look all around the inside of their shells too. *What do you see? Do all of the insides of the shells look the same? Are there noticeable differences?*

EXTRA

Do fresh eggs and aged eggs behave similarly? Buy a dozen eggs whose expiration date is at least two weeks away. Try this activity with half of the eggs right away. Let the other six eggs age in the refrigerator for two weeks. Repeat the activity with the aged eggs. *How does the data compare between the fresh and the aged eggs?*

EXTRA

If pores in the chicken egg's shell allow materials to cross back and forth between the inside of the egg and the outside environment, then the air inside the egg could be replaced by water, and water is heavier than air. Using a scale that can distinguish changes as small as 0.1 gram, such as a triple-beam balance or high-quality electronic kitchen scale, weigh some eggs, then have an adult help you hard-boil them and weigh the eggs again. *Did the eggs change weight? If so, how did they change weight? What does this say about the ability of the chicken egg to allow water to cross its shell?*

OBSERVATIONS AND RESULTS

Did all of the eggs have at least a few small blue dots on the inside of their shells? Were the dots mostly clustered in one or a few areas on the inside of each shell?

Directly under a chicken egg's shell are two membranes. When the eggs are laid by the mother, they are warmer than the air, and as they cool the material inside the egg shrinks a little bit. This shrinkage is what pulls the two membranes apart, leaving behind the small air sack that is filled with oxygen. As the developing chick grows, it uses the oxygen from the air sack and replaces it with carbon dioxide. The tiny pores in the shell allow the carbon dioxide to escape and fresh air to get in. A chicken egg has more than 7,000 pores in its shell to allow this to happen! These pores also allow water to go through the shell, which is why the dye appears as small dots on the inside of the shell, often clustered in certain areas, and why an egg after being hard-boiled would weigh slightly more than when it was raw. Also, freshly laid eggs do not allow water to penetrate as well as older, commercial eggs do, so fewer blue spots will probably be visible on the inside of fresher eggs compared with older ones.

CLEANUP

Dispose of the raw eggs by pouring them down the drain. (The eggs should not be eaten because they were soaked with dish detergent.) Thoroughly clean any surface the raw eggs touched because they can carry salmonella.

How Animals Stay Warm with Blubber

WOULD BLUBBER HELP YOU STAY WARMER LONGER? TRY OUT THIS ACTIVITY!

PROJECT TIME

10 to 30 minutes

KEY CONCEPTS

Temperature
Heat transfer
Adaptations
Insulation
Fatty tissue

Have you ever wondered how whales and other marine mammals survive and keep warm in the cold oceans? Warm-blooded mammals can live in these chilly conditions because their bodies have some cool warmth-saving adaptations, thanks to generations of natural selection. In other words, to pass on characteristics (via their genes), the predecessors of modern marine mammals had to overcome different challenges to reproduce, and their descendants received the genes that allowed for their survival. This kind of change in organisms over time is what fuels evolution. An important adaptation for marine mammals is blubber, a thick, insulating layer of fat beneath the skin that helps to keep body warmth in and the cold of the air or water out. Will a layer of fake blubber—in the form of shortening—help you keep from getting cold?

BACKGROUND

Mammals that have evolved to live in cold waters, such as whales, seals, sea lions, and polar bears, commonly have a layer of blubber. Whether they are living in cold waters near the North Pole or around Antarctica or are visiting the deep ocean, these animals' blubber is vital to their survival. During the winter, the air in the Arctic (the northernmost part of the world) is often below -40°F (-40°C). Antarctica, the coldest place in the world, can be below -76°F (-60°C). Depending on the species, whales dive more than 1/4-mile (400 m) deep in the ocean, where the water can be colder than 54°F (12°C).

Blubber helps these marine mammals from getting too cold. (Cold-blooded marine animals, such as fish, sharks or crabs, do not need to stay warm and can let their body temperatures get closer to that of the water. Thus, they do not need to have this extra insulation.) Blubber is a thick layer of fat (adipose) tissue. Animals store extra digested food in the form of adipose tissue, which contains molecules called lipids. Adipose tissue has a relatively low thermal conductivity, which means that it does not transfer heat as well as other tissues and materials—such as muscle or skin. That way, it helps to insulate an animal's body.

MATERIALS

- Two bowls
- Cold water
- Warm water
- Ice cubes
- Shortening (such as Crisco)
- Paper towels
- Stopwatch
- Thermometer
- A partner

PREPARATION

- Put an equal number of ice cubes into each bowl without filling either bowl too full. Add cold water to each bowl.
- Measure the temperature of the water in each bowl with a thermometer. They should be the same temperature. When the temperature levels off (which should happen quickly), the water is ready for the test.

PROCEDURE

- Cover your pointer finger on one hand with a thick layer of shortening, covering the entire area that will be submerged in the water. Leave your other pointer finger clean and bare.
- Have your partner prepare the stopwatch. When they are ready, put the pointer finger of each hand into one of the bowls of ice water and have your partner start timing you. As soon as your finger feels too cold to keep it in the water any longer, take it out. *How long did you leave each finger in the bowl?*
- Let your fingers warm up and return to their normal color. If any shortening came off the covered finger, reapply it.
- Have your partner help you pour the cold water down the sink and refill the two bowls with warm water (make sure it is warm but not hot enough to burn the skin).
- Measure the temperature of the water in each bowl with a thermometer. They should be about the same. *In the warm water, do you think you'll see the same result?*
- Have your partner time how long you can leave each finger in the bowls of warm water. *How long did you leave each finger in the bowl? Was the time difference between the two fingers larger or smaller than when you put your fingers in the cold water?*

EXTRA

How consistent are your results? You can repeat this activity two or three times, recording the temperature of the different waters tested and the time each finger was in the water. Then make a graph out of your results. *In which environment did the shortening "adaptation" consistently give an advantage?*

OBSERVATIONS AND RESULTS

When you dunked your fingers in the ice-cold water, did the finger covered in shortening stay warm longer than the finger that was not covered? Was there not as large a time difference when you put your fingers in the warm water?

Mammals that have adapted to live in cold waters—such as polar bears and whales—can stay warm largely because of their blubber, a thick layer of fat. The blubber is evenly spread over much of their body, just as the shortening in this activity covered the surface of your finger in a thick layer. Since adipose tissue has a relatively low thermal conductivity, it does not transfer heat well compared to other tissues and materials. Humans have developed some insulating materials for our own daily use. For example, Styrofoam is another material that does not conduct heat well, whereas metals conduct heat very well. This is why hot drinks are often served in Styrofoam, since it keeps the heat inside the cup, thus preventing your hands from being burned. (For the same reason, it can also keep cold drinks cool longer than, for example, a paper cup.) What other materials can you think of that work as insulators?

CLEANUP

Wipe the shortening off your hands and then wash them with soap and water. Pour leftover water down the drain. Put away your other materials where you found them.

THE SCIENTIFIC METHOD

The scientific method helps scientists—and students—gather facts to prove whether an idea is true. Using this method, scientists come up with ideas and then test those ideas by observing facts and drawing conclusions. You can use the scientific method to develop and test your own ideas!

Question: What do you want to learn? What problem needs to be solved? Be as specific as possible.
Research: Learn more about your topic and refine your question.
Hypothesis: Form an educated guess about what you think will answer your question. This allows you to make a prediction you can test.
Experiment: Create a test to learn if your hypothesis is correct. Limit the number of variables, or elements of the experiment that could change.
Analysis: Record your observations about the progress and results of your experiment. Then analyze your data to understand what it means.
Conclusion: Review all your data. Did the results of the experiment match the prediction? If so, your hypothesis was correct. If not, your hypothesis may need to be changed.

GLOSSARY

aviary: A place, such as a large cage, where many birds are kept.

binoculars: A tool to look through to see things far away.

biodiversity: The different kinds of life in an environment shown by numbers of different kinds of plants and animals.

biologist: A scientist who studies living things, such as animals.

classify: To assign something to a category or class based on shared qualities.

commercial: Widely sold by a company.

conduction: The movement of heat through something.

convection: The movement of a gas or liquid when the warmer parts go up and the colder parts go down. Also, the movement of heat by this process.

field guide: A book that helps identify certain animals or plants in an area.

habitat: The place or type of place where a plant or animal naturally or normally lives or grows.

radiation: The giving off of energy, such as in the form of heat, through waves that cannot be seen.

respiration: The act of breathing.

species: A group of plants or animals that are all of the same kind.

taxonomic: Having to do with taxonomy, or the way living things are put into groups based on shared features.

theory: An explanation based on facts that is generally accepted by scientists.

ADDITIONAL RESOURCES

Books

Heinecke, Liz Lee. *The Kitchen Pantry Scientist Biology for Kids: Science Experiments and Activities Inspired by Awesome Biologists, Past and Present.* Glouster, MA: Quarry Books, 2021.

Olson, Elsie. *Not-So-Ordinary Science: 49 Projects that Ooze, Pop, Zoom, and More!*. North Mankato, MN: Dabble Lab/Capstone Press, an imprint of Capstone, 2023.

Pottle, Jules. *1,000 Words: Nature*. London, UK: Dorling Kindersley Limited, 2022.

Websites

Discovery Education
sciencefaircentral.com

Exploratorium
www.exploratorium.edu/search/science%20fair%20projects

Science Buddies
www.sciencebuddies.org/science-fair-projects/project-ideas/list

Science Fun for Everyone!
www.sciencefun.org/?s=science+fair

Videos

Adaptations of Arctic Animals
ny.pbslearningmedia.org/resource/nat15.sci.lisci.arctanim/adaptations-of-arctic-animals/

Animal Adaptations: Intelligence and Social Behavior
ny.pbslearningmedia.org/resource/nat37-sci-puzzle/animal-adaptations-intelligence-and-social-behavior/

How Animals Affect Human Health
ny.pbslearningmedia.org/resource/how-animals-affect-human-health-video/ecosense-for-living/

Publisher's note to educators and parents: Our editors have carefully reviewed these websites to ensure that they are suitable for students. Many websites change frequently, however, and we cannot guarantee that a site's future contents will continue to meet our high standards of quality and educational value. Be advised that students should be closely supervised whenever they access the internet.

INDEX